Placer Gold Deposits of Utah

By MAUREEN G. JOHNSON

GEOLOGICAL SURVEY BULLETIN 1357

A catalog of location, geology, and production with lists of annotated references pertaining to the placer districts

UNITED STATES GOVERNMENT PRINTING OFFICE, WASHINGTON : 1973

UNITED STATES DEPARTMENT OF THE INTERIOR

ROGERS C. B. MORTON, *Secretary*

GEOLOGICAL SURVEY

V. E. McKelvey, *Director*

Library of Congress catalog-card No. 72-600159

For sale by the Superintendent of Documents, U.S. Government Printing Office
Washington, D.C. 20402 - Price $1.25 (paper cover)
Stock Number 2401-00264

CONTENTS

	Page
Abstract	1
Introduction	1
History of placer mining in Utah	1
Purpose and scope of present study	3
Garfield County	4
1. Colorado River placers	4
2. Imperial district (Crescent Creek placer)	5
Grand County	6
3. Colorado River placers (Grand River)	6
4. La Sal district (includes Miners Basin)	7
Millard County	9
5. House Range placers (Sawtooth district, Notch Peak placers)	9
Salt Lake County	10
6. Bingham district	10
San Juan County	12
7. San Juan River placers	12
8. Lower Colorado River placers	13
9. Blue Mountain district	14
Uintah County	15
10. Green River placers	15
Other placer districts	16
Emery County	16
11. Emery district	16
Juab County	16
12. Detroit (Drum) district	16
Piute County	17
13. Gold Mountain district	17
14. Ohio (Mount Baldy) district	17
Sevier County	17
15. District unknown	17
Tooele County	17
16. Camp Floyd (Mercur) district	17
17. Clifton (Gold Hill) district	18
Wayne County	18
18. Colorado River placers	18
Gold production from placer deposits	18
Summary	20
Placers associated with known lode deposits	20
River placers	21
Bibliography	22
Literature references	22
Geologic map references	25

CONTENTS
ILLUSTRATION

PLATE 1. Map of placer deposits of Utah..............................In pocket

TABLE

Page

TABLE 1. Utah placer gold production................................... 19

PLACER GOLD DEPOSITS OF UTAH

By Maureen G. Johnson

ABSTRACT

Eighteen placer districts in Utah are estimated to have produced a minimum of 85,000 ounces of placer gold from 1864 to 1968. Summary of the location, areal extent, past production, mining history, and probable lode source for each district is based on information obtained from a wide variety of published reports relating to placer deposits. Annotated references to all reports give information about individual deposits for each district.

The most important placer district in Utah is the Bingham district, which produced more than 75,000 ounces of placer gold. The placers in the Bingham district occurred in thick gravel deposits in Bingham Canyon and tributary drainages and were derived mostly from oxidized copper and lead-zinc-silver ores of Oligocene age. Most of the other productive placers in Utah occur along the major river drainages in the eastern part of the State, where sporadic accumulations of flake gold in sand and gravel bars have been mined. Small placers are found associated with lode deposits in scattered areas throughout Utah.

INTRODUCTION

HISTORY OF PLACER MINING IN UTAH

Placer mining in Utah began in 1864 at Bingham Canyon, Salt Lake County. One minor deposit was reportedly found before this discovery—at Gold Hill in the Deep Creek Mountains (Clifton district, Tooele County) in 1858—but Indians are said to have driven away the early prospectors here. The first permanent settlers in Utah were Mormons who, under the leadership of Brigham Young, arrived in 1847. The Mormons were interested in permanent self-sufficient settlements and concentrated their energies in developing the land for ranching and farming. Prospecting Utah began in earnest when General P. E. Conner and troops of California volunteers established a base at Camp Douglas, near Salt Lake City. These soldiers, many of whom were experienced prospectors and miners, quickly began investigating the surrounding area for metalliferous deposits. OnSeptember 17, 1863, the first mineral location in Utah (the West Jordan lode claim) was made, and the placers were discovered in the following year (Butler and others, 1920).

By 1865 the placers in Bingham Canyon were being extensively worked;

production from the gravels reached a peak between 1868 and 1872, when placer gold valued at about $1 million was recovered. Placer mining gradually decreased in importance after 1872 as the richer gravels were worked out and was negligible by 1900. The beginning of open-pit mining of copper ores containing small amounts of gold, about 1907, and continual expansion of the open-pit mine in the following years have resulted in the virtual obliteration of the original topography and many of the remnants of placer gravels. Doubtless, some workable placers were destroyed by the copper mining operations, but these deposits probably were not so valuable as those worked in the early years.

Placer gold was discovered in other lode mining districts in the State before 1900, but none of these deposits, such as those reported in the Gold Mountain and Ohio districts (Piute County) and Camp Floyd district (Tooele County), were of economic importance.

After the decline of placer mining in the Bingham district, placer deposits along the Colorado, San Juan, and Green Rivers in the eastern part of the State were extensively investigated by many itinerant prospectors. The placers were found in sand and gravel bars along many miles of these river courses; the gold, however, was exceedingly fine and difficult or impossible to recover. In 1891 a major placer rush to the San Juan River began. Placers had been worked along the river desultorily since 1879, but exaggerated reports and rumors of rich placers, spread by a trader named "John Williams," caused an influx of men to this relatively isolated area during the "Bluff excitement of 1892." The majority of the men returned from the San Juan River placers within a few months, empty handed and penniless. Despite the quick end of the placer rush to the San Juan River, the presumed occurrence of large quantities of placer gold along the San Juan and Colorado Rivers attracted the attention of mining companies and men who invested large amounts of time, energy, and money to develop the placers. Factual information about the success or failure of many of these enterprises is difficult to obtain, but placer mining at Good Hope Bar and California Bar on the Colorado River was said to be successful, whereas a dredging operation in 1900 on the Colorado River near New Year Bar was a failure.

During the 20th century, placer mining has continued sporadically in a few districts in the State. The placers in the La Sal Mountains, discovered in 1907, created a temporary interest in placer mining in that area, but low gold recovery disappointed the miners. The placers discovered most recently in Utah are those located in the House Range (Millard County), first worked in 1932.

Construction of the Glen Canyon Dam has resulted in the gradual rise of the level of the Colorado River since 1963, and the rising waters of Lake Powell will eventually drown many of the gold-bearing sand bars along the Colorado and San Juan Rivers.

INTRODUCTION

PURPOSE AND SCOPE OF PRESENT STUDY

The present paper is a compilation of published information relating to the placer gold deposits of Utah, one of a series of four papers describing the gold placer deposits in the Southwestern States. The purpose of the paper is to outline areas of placer deposits in the State of Utah and to serve as a guide to their location, extent, production history, and source. This work was undertaken as part of the investigation of the distribution of known gold occurrences in the Western United States.

Each placer is briefly described. Location is given by geographic area and township and range. Topographic and geologic maps which show the placer area are listed. Access to each area is indicated by direction and distance along major roads and highways from a nearby center of population.

Detailed information relating to the exact location of placer deposits, their thickness, distribution, and average gold content (all values cited in the text have been converted to gold at $35 per ounce, except where otherwise noted) is included, where available, under "Extent." U.S. Bureau of Land Management land plats of surveyed township and ranges were especially helpful in locating old placer claims and some creeks and drywashes not named on recent topographic maps. These land plats were consulted for all the surveyed areas in Utah. U.S. Bureau of Mines records were also consulted for the location of small placer claims.

Map publications of the Geological Survey can be ordered from the U.S Geological Survey, Distribution Section, Denver Federal Center, Denver, Colo. 80225; book publications, from the Superintendent of Documents, Government Printing Office, Washington, D.C. 20402.

Discovery of placer gold and subsequent placer-mining activity are briefly described under "Production history." Detailed discussion of mining operations is omitted, as this information can be found in the individual papers published by the State of Utah, in the yearly Mineral Resources and the Mineral Yearbook volumes published by the U.S. Bureau of Mines and the U.S. Geological Survey, and in many mining journals. Placer gold production, in ounces (table 1), was compiled from the yearly Mineral Resources and Mineral Yearbook volumes and from information supplied by the U.S. Bureau of Mines. These totals of recorded production are probably lower than actual gold production, for substantial amounts of coarse placer gold commonly sold by individuals to jewelers and specimen buyers are not reported to the U.S. Bureau of Mines or to the U.S. Bureau of the Mint. Information about the age and type of lode deposit that was the source of the placer gold is discussed for each district.

A detailed search of the geologic and mining literature was made for information concerning all placers. A list of literature references is given with each district; the annotation indicates the type of information found in each reference. Sources of information are detailed reports on mining districts, general geologic reports, Federal and State publications, and brief articles and news notes in mining journals. A complete bibliography, given at the end of the paper, includes separate sections for all literature references and all geologic map references.

GARFIELD COUNTY

1. COLORADO RIVER PLACERS

Location: Along the Glen Canyon part of the Colorado River from the mouth of Dirty Devil River south to Lake Canyon, T. 34 S., R. 13 E. to T. 39 S., R. 11 E.

Topographic maps: All 15-minute quadrangles—Hite, Mancos Mesa, Mount Ellsworth, Lake Canyon.

Geologic map: Hunt, Averitt, and Miller, 1953, Geologic map of the Henry Mountains (pl. 1), scale 1:125,000.

Access: From Hanksville, it is about 45 miles southeast on State Highway 95 to the Colorado River; gravel bars and terraces along the river edge are accessible by trail and boat. The area is being developed as part of the Glen Canyon National Recreation area and administered by the U.S. National Park Service.

Extent: Placers are found at gravel bars along the Colorado River in Garfield County (on the west bank of the river) and in San Juan County (on the east bank of the river). Most of the placers occur in terrace gravels that were 175 feet or less above the river before completion of the Glen Canyon Dam in 1963. The rising waters of Lake Powell will eventually reach an elevation of 3,700 feet and drown most, if not all, placer deposits. Before construction of the dam, the height of the river at Hite's Crossing was about 3,500 feet, and at Lake Canyon, about 3,300 feet.

The gold generally occurs in small flakes about 0.10 mm in diameter; some flakes as large as 0.36 mm long and 0.26 mm wide were found at the Gold Coin placer (Olympia Bar, San Juan County, Mount Ellsworth quadrangle). The gold is either distributed throughout the gravels or is concentrated near the surface layer of the gravels and at the gravel-bedrock contact. As the gold content in the gravels varies widely, reports of average value per cubic yard have little meaning.

Production history: Gold was first discovered in Glen Canyon in 1883 by Cas Hite, who discovered placer gold at Dandy Crossing 1 year after Carl Shirts prospected for placer gold at Burro Bar farther downstream. Within a few years, placer gold was found at numerous bars along the river, and placer-mining operations began at several places along the river. Most activity occurred between 1886-89, declining thereafter. Production from the Colorado River placers was never very large because of the difficulty in recovering gold that occurred as very fine flakes. Part of the production was credited to San Juan County; some of the better known placer localities are indicated by letter on plate 1: a, White Canyon; b, Red Canyon; c, Tickaboo Bar; d, Good Hope Bar; e, California Bar; f, New Year Bar. California Bar reportedly produced placer gold valued at about $10,000, apparently before 1900; White Canyon, Red Canyon, and Tickaboo Bar have been mined on a small scale during the 20th century. In 1900 the Hoskaninni Co. installed a dredge in the river near New Year Bar in an unsuccessful attempt to mine the river gravels.

Source: Unknown. The fine size of the gold and the presence of gold along other parts of the Colorado River upstream from Glen Canyon indicate a distant upstream source. Gregory (1938, p. 108), however, suggests that the gold was derived from erosion of the Triassic and Jurassic rocks exposed in the area.

Literature:

Butler and others, 1920: Describes placers at Gold Coin deposits; size of gold; accessory minerals; distribution; average value.

Gregory, 1938: Early history; placer mining in 1927; size of gold; accessory minerals; source.

Gregory and Moore, 1931: History; location of placer-mining claims and prospects; production of California Bar; size and distribution of gold in the gravels; gold values per cubic yard; placer-mining operations.

Hunt and others, 1953: Summarizes placer operations in Glen Canyon; names bars worked for placer gold; gives height above river; history of placer-mining operations; history of placer discovery.

Mining Reporter, 1899: Brief, but exaggerated, note on wealth of placer gravels of Colorado River in vicinity of Dandy's Crossing.

Thaden and others, 1964: History; lack of placer-mining activity.

2. IMPERIAL DISTRICT (CRESCENT CREEK PLACER)

Location: East flank of Mount Ellen in the Henry Mountains, T. 31 S., R. 11 E.

Topographic map: Bull Mountain 15-minute quadrangle.

Geologic map: Hunt, Averitt, and Miller, 1953, Geologic map of Mount Ellen and Mount Pennell (pl. 7), scale 1:31,680.

Access: The placers are about 5 miles west of State Highway 95, about 24 miles south of Hanksville, and are easily accessible by dirt roads leading from the State highway.

Extent: Placers are found in the gravels of Crescent Creek, an eastward draining tributary of North Wash, which is, in turn, a tributary to the Colorado River. Crescent Creek heads at Bromide Basin at the crest of Mount Ellen and is the only creek in the region known to contain economic concentrations of placer gold. The gold at the Lawler-Ekker placer, about 5 miles east of Bromide Basin (sec. 28, T. 31 S., R. 11 E., Bull Mountain quadrangle), is concentrated in black-sand streaks at the base of the fanglomerate gravel in benches along Crescent Creek; most of the gold is about 0.5 mm in diameter, but flakes as much as 2 mm long and 1 mm thick are found.

Noneconomic deposits of placer gold are also reported in Straight Creek, which drains the east flank of Mount Pennell (T. 33 S., Rs. 10 and 11 E.) about 7 miles south of Mount Ellen. Only very small amounts of gold have been reported from creek gravels on the north and west flanks of Mount Ellen and in North Wash and Hansen Creek.

Production history: Gold was found in the Mount Ellen area in 1890, but placer gold apparently was mined in 1888 near Mount Pennell. Accurate production information for the early years of placer mining in the Henry Mountains is difficult to find. In 1893, men were reportedly recovering $5-$10 per day despite a shortage of water, and one man, J. F. Wilson, reportedly recovered $2,700 in 4 months with the use of drywashers. At the Lawler-Ekker placer, which reportedly produced a few thousand dollars, the value of the gravel was reported at 50¢ to 75¢ per cubic yard. Other placers worked in the area include the Emery, Burro, Big Bend, Eagle City, and North Wash claims.

Source: Hunt, Averitt, and Miller (1953) state that the placers are confined to stream courses that drain the stocks in the Henry Mountains. Only Crescent Creek, of the streams draining Mount Ellen, is known to contain valuable placers; the gold concentrated in the stream was derived from gold-copper fissure veins at Bromide Basin in the Mount Ellen stock (predominantly diorite porphyry).

Small irregular veinlets containing gold near the south edge of the central monzonite porphyry core of the Mount Pennell stock on the north side of Straight Creek were the likely source of placer gold in that creek.

Literature:

Butler and others, 1920: Extent; production; source; accessory minerals.

Engineering and Mining Journal, 1897a: Reports placer activity near Mount Pennell for the period 1888-91.

———— 1893a: Reports production per day per man on placers at Henry Mountains; amount of gold recovered by J. F. Wilson with drywashing machines; developments at placer properties; sale price of some placer claims.

Hunt and others, 1953: Location of gold placer property; size and distribution of gold; accessory minerals; grades of gravel mined; production; source.

Mining Journal, 1929: Placer-mining activity; value of gold in concentrate; Henry Mountains.

GRAND COUNTY
3. COLORADO RIVER PLACERS (GRAND RIVER)

Location: Along the Colorado River from the mouth of the Dolores River downstream to the Amasa Back bend in the river west of Moab, especially T. 23 S., Rs. 23 and 24 E.; T. 24 S., R. 23 E.; T. 26 S., R. 21 E.

Topographic maps: All 15-minute quadrangles—Cisco, Castle Valley, Moab.

Geologic map: Williams, 1964, Geology, structure, and uranium deposits of the Moab quadrangle, scale 1:250,000.

Access: State Highway 128 parallels the Colorado River north of Moab; State Highway 279 parallels the river west and south of Moab.

Extent: Placers have been found at various places along the Colorado River in Grand County. The deposits occur near the mouth of the Dolores River (secs. 9 and 16, T. 23 S., R. 24 E., Cisco quadrangle), near Hittle Bottom in Professor Valley (sec. 35, T. 23 S., R. 23 E., Cisco quadrangle), in Professor Valley near Richardson (sec. 16, or 20, T. 24 S., R. 23 E., Castle Valley quadrangle) and at Gold Bar west of Moab (sec. 5, T. 26 S., R. 21 E., Moab quadrangle). Placers have also been reported elsewhere along the Colorado River, in particular at Hamlin Bar and the Rio Grande Group, but the location of these has not been adequately described.

No descriptions have been found of the thickness of the gravels or the distribution of gold in the gravels.

Production history: The placers along this part of the Colorado River were worked almost continually until 1942, and, although yearly production of placer gold was never large, the total placer gold production during this century is second only to the Green River in the State. The most actively worked placer was Hamlin Bar, which apparently is at, or near, Dewey (sec. 18, T. 23 S., R. 24 E., Cisco quadrangle); this deposit was worked periodically between 1906 and 1928 and was dredged during 1927–28.

A few ounces of placer gold credited to the Dolores River are included with the placer production from the Colorado River in Grand County, because this gold was probably recovered from gravels in this river at, or near, its junction with the Colorado.

Source: Unknown. The gold occurs as fine particles, similar to the gold in the Glen Canyon placers; it is difficult to recover, and probably was derived from a bedrock source an unknown distance upstream.

Literature:
  Dane, 1935: Placer-mining activity near Dewey, 1927–28.
  Engineering and Mining Journal, 1936: Location of placer claims.
  Ritzma and Doelling, 1969: Sketch map shows location of two placer deposits.

4. LA SAL DISTRICT (INCLUDES MINERS BASIN)

Location: Wilson and Bald Mesas, Miners Basin, west and south of North Mountain in the La Sal Mountains. (Partly within the Manti-La Sal National Forest.) T. 26 S., Rs. 23 and 24 E.

Topographic map: Castle Valley 15-minute quadrangle.

Geologic maps:
>Hunt, 1958, Geologic map of the North La Sal stock (pl. 40), scale 1:20,000.
>
>Richmond, 1962, Geologic map of the Quaternary deposits of the La Sal Mountains (pl. 1), scale 1:48,000.

Access: From Moab, it is about 16 miles northeast on State Highway 128 to the road leading southeast through Castle Valley. Dirt roads lead south from the Castle Valley road about 7 miles southeast of State Highway 128 to placer areas.

Extent: Gold occurs in glacial deposits on the mesas southwest of North La Sal Mountain and south of Castle Valley and in Miners Basin and Placer Creek on the west flank of North La Sal Mountain at the western edge of Castle Valley. The placers on Wilson Mesa (approximately secs. 10 and 15, T. 26 S., R. 23 E.) and Bald Mesa (approximately sec. 19, T. 26 S., R. 24 E.) occur in Pleistocene pre-Wisconsin deeply weathered glacial gravels, which vary from thin layers to deposits 50 feet or more thick. The gravels are composed of subangular debris (mostly soda syenite, feldspathoidal rocks, metadiorite, and amethystine quartz) derived from the central part of North La Sal Mountain. The gold is disseminated throughout the gravels and occurs as small wires or flakes that do not appear to be waterworn. The placers in Placer Creek, east of Pinhook (approximately sec. 16, T. 26 S., R. 24 E.), and in the upper part of Miners Basin (approximately sec. 15, T. 26 S., R. 24 E.) are found in little-weathered glacial outwash and moraine; the gold is fine and sparsely distributed.

Production history: The placers on Wilson Mesa were discovered in 1907 and were worked intermittently until 1948. Most activity took place during the first few years after discovery of the gold; sluicing was the most common mining technique used. For many years the placer production from Miners Basin and that from Wilson Mesa were reported separately; judged by these records, the Wilson Mesa placers were the most productive.

Source: The placer gold was derived from fissure deposits in the North La Sal Stock (predominately diorite porphyry with a central core of metadiorite and soda syenite); the gold occurs in pyrite in quartz veins along the fissures and in altered wallrock. Where the pyrite has been oxidized, free gold can be panned. According to Hunt (1958, p. 355), the deeply weathered gravels offer more promise for higher gold concentrations than the later less-weathered gravels, because gold contained in sulfide minerals in the igneous rocks in the gravels would be freed by long periods of oxidization.

Literature:
>Butler and others, 1920: Quotes Hill's description of placers (1913); discusses origin of the gravels.

Hill, 1913: Extent of placer gravels; thickness; size of debris in gravels; lithology of gravels; fineness and size of gold; placer-mining techniques; production; source.

Hunt, 1958: Location; character of placer gravel; source. Areas suggested for prospecting.

Mining Review, 1910b: Reports value of placer gravels at 25¢ to 87½¢ per yard (at $20.67 per yard); production in 5 months by two men, $1,900.

Mining Science, 1910: Names creeks where placer gold was found; reports values of 25¢ to $1 per yard (at $20.67 per yard) in Wilson Mesa; depth of gravels.

MILLARD COUNTY
5. HOUSE RANGE PLACERS (SAWTOOTH DISTRICT, NOTCH PEAK PLACERS)

Location: Central part of the House Range in Amasa Valley and Granite Canyon, T. 19 S., R. 14 W.

Topographic map: Notch Peak 15-minute quadrangle.

Geologic map: Hanks, 1962, Geology of the central House Range area, scale about 1 in. = 1 mile.

Access: The placer area lies within the part of the House Range between U.S. Highway 6-50 and old U.S. Highway 50, about 44 miles west-southwest of Delta. Amasa Valley and Granite Canyon are accessible by dirt roads branching off these highways.

Extent: Placer gold has been recovered from the arkosic sands and gravels in Amasa Valley (sec. 2, T. 19 S., R. 14 W.) and from gravels in Granite Canyon near the point where the north and east forks of the creek join the main canyon bed (sec. 12, T. 19 S., R. 14 W., projected). Granite Canyon is named "Miller Canyon" on the topographic map of the area.

Production history: Placer production in the House Range was first recorded in 1932 and continued yearly until 1943 and intermittently until 1957. The production was credited to the Sawtooth district from 1932 to 1940 and in 1957, and to the House Mountains, from 1941 to 1953. Most of the placer gold was recovered from claims in the Amasa Valley, but gold was recovered from Granite Canyon in 1940. During the late 1930's, the placers were investigated by the Mineral Valley Gold Mining Co., but large-scale placer mining was never attempted.

Source: The source of the gold in the placers is uncertain. The arkosic sands and gravels in which the placers are found were derived by weathering of the quartz monzonite stock that forms Notch Peak. Crawford and Buranek state that the gold was derived from quartz veins, but Gehman (1958, p. 46) points out that the only quartz observed was found in pegmatites and scheelite veins barren of gold.

Literature:

Crawford and Buranek, 1944: Source of placer gold in Granite Canyon.

Gehman, 1958: Lithology of placer gravels; source of gold in Amasa Valley.
Hanks, 1962: Placer-mining production.
Mining Journal, 1938: Reports gold, mercury, tungsten, and molybdenum in gravels as much as 30 feet thick.
Powell, 1959: Notes presence of placer claims in Granite Canyon.

SALT LAKE COUNTY
6. BINGHAM DISTRICT

Location: In the Oquirrh Mountains, in and adjacent to the Bingham open-pit copper mine, T. 3 S., R. 3 W.

Topographic maps: Bingham Canyon and Lark 7½-minute quadrangles. Bingham mining map, special edition, 1901, scale 1:20,000 (for location of gulches before open-pit mining).

Geologic maps:
Boutwell, 1905, Map of economic geology of Bingham mining district and index to mines, pl. 16 (shows distribution of placer gold deposits), scale 1:20,000.
Utah Geological Society, 1961, Kennecott Copper Corporation surface geology map of the Bingham district (pl. 2), scale ½ in. = 1,000 ft.

Access: From Salt Lake City, 12 miles south on U.S. Interstate 15 to Midvale; 14 miles west on State Highway 48 to Bingham.

Extent: Placers were found in Bingham Canyon from its upper branches—Carr Fork, upper Bingham Canyon, Bear Gulch—downstream along the main canyon to the mouth of Bingham Canyon, where the creek enters Jordan Valley. Most of the original topography has been obliterated by the open-pit mine and waste dumps outside the mine. Placer gold was found in bench and creek gravels (as much as 250 ft thick) which represent deposition, reworking, and redeposition of eroded material at successively later periods of dissection of Bingham Canyon. Most of the gold values were found at one of two horizons—the gravels immediately overlying bedrock, and the "upper leads," where the gold is concentrated in paystreaks above dense gravel beds forming a false bedrock. The gravels consist of the more resistant rock types exposed in the area (predominantly quartzites and porphyries) and become progressively more rounded downstream; the gold contained in the gravels becomes finer and more rounded downstream. The gold content of the gravels varied considerably in different deposits—sampling at the Argonaut pit (about 1900) indicated an average of 10¢ per yard for the lower 30 feet of bench gravels, a yield of $31 per cubic yard of gravel (probably from a rich pocket over bedrock) at the Old Channel deposit.

Production History: The Bingham placers were discovered in 1864, 1 year after the discovery of lead ore. Historically, the placers at Bingham were the largest in Utah in production and extent; in the 20th century, the placers at Bingham have had negligible production. About $1 million in placer gold was recovered from the gravels by 1871, and another $500,000 by the turn of the century. The greatest period of placer-mining activity was from 1868 to 1872; after that time, placer mining declined in importance relative to lode mining. Most of the early work utilized sluicing, drift mining, and hydraulic mining techniques. Many of the placer deposits were considered to be worked out by the late 1800's, but Boutwell (1905, p. 377) indicates that in 1900 some workable deposits probably remained at Dixon Bar, Bear Gulch, and buried creek gravels in lower Bingham Canyon. Small amounts of placer gold were recovered from the district during the period 1920–21 and again during the period 1932–35.

Source: Gold occurs in minor amounts in the porphyry-type copper ore, and in the "nonporphyry" copper and lead-zinc-silver ores. In some of the oxidized parts of the nonporphyry ores, gold has been concentrated in honeycombed siliceous gangue, and rapid erosion of these deposits provided the source for most of the placer gold in the district. Boutwell (1905, p. 342) points out that some of the placer gold in upper Bear Gulch does not seem to have been derived from oxidized ore and probably was derived from mineralized porphyry. The relation of intrusion of the Bingham stock to the alteration and subsequent mineralization in the district has been studied in detail by many geologists. Moore, Lanphere, and Obradovich (1968) date the intrusion of the Bingham stock at 37.7–37.2 m.y. (million years) and the age of alteration at 36.5 m.y (average); thus, the ore minerals in the Bingham district, some of which were the source of placer gold, formed about 36.5 m.y. ago.

Literature:

Boutwell, 1902: Abstract of U.S. Geological Survey Prof. Paper 38.

——— 1905: Detailed and complete description of placer deposits. Includes map showing location of placers; description of gravels; gold content; past placer-mining operations; erosional history of Bingham area; source of placer gold; problems encountered in placer mining; future potential in placer mining.

Burchard, 1882: Production statistics for 1881.

Butler and others, 1920: Quotes Boutwell's description (1905) of the placers.

Hammond, 1961: History.

Maguire, 1899: History.

Moore and others, 1968: Radiometric ages of altered and unaltered igneous rocks; age of mineralization.

Raymond, 1870: Early small-scale placer-mining operations in 1869.

——— 1872: Placer-mining operations in 1870.

——— 1873: Placer-mining activity in 1871; production.

——— 1874: Placer-mining activity in 1873; production.

———— 1877: Placer-mining activity in 1875; production.

Rubright and Hart, 1968: Ore mineralogy; gold content in nonporphyry ores; source of placer gold.

SAN JUAN COUNTY
7. SAN JUAN RIVER PLACERS

Location: Along the San Juan River from the mouth of Montezuma Creek west to the junction with the Colorado River, Tps. 40 and 41 S., Rs. 9–23 E.

Topographic maps: All 15-minute quadrangles—Montezuma Creek, Mexican Hat, Clay Hills, No Mana Mesa.

Geologic map: Cooley, Harshbarger, Akers, and Hardt, 1969, Geologic map of the Navajo and Hopi Indian Reservations (pl. 1, sheets 3, 5), scale 1:125,000.

Access: The placers along the San Juan River are difficult to reach. Placers east of Mexican Hat are accessible from State Highways 47 and 262, which parallel the San Juan River. Placers west of Mexican Hat are accessible by a dirt road that branches off State Highway 95 toward Clay Hill Crossing. Parts of the river are accessible by trail or boat.

Extent: Gold-bearing gravel deposits are found at different localities along the San Juan River from the mouth of Montezuma Creek west to the junction with the Colorado River. The deposits are in gravels at river level to about 100 feet above the level of the river at the time of completion of the Glen Canyon Dam in 1963. The rising waters of the San Juan River arm of Lake Powell will extend up the river to Piute Farms, and will eventually reach a high level elevation of about 3,800 feet. Most, if not all, of the placer gravels west of Piute Farms (30 miles west of Mexican Hat), including the gravels at Spencer and Zahn's Camp, will be inundated. A few gravel deposits east of Piute Farms, including those at Honaker Trail, will probably not be affected by the elevation of the waters of the San Juan River. Some of the well-known placer localities along the river are indicated by letter on plate 1 as follows: a, placers below the mouth of Montezuma Creek; b, placers near Honaker Trails; c, Piute Farms; d, Clay Gulch; e, Spencer and Zahn's Camp.

Production history: Placer gold apparently was discovered in San Juan River gravels in 1879, and prospecting continued intermittently until 1891. Reports of rich placer deposits in the area were circulated during the period 1891–92, and reportedly more than 1,000 men rushed to the San Juan River during the "Bluff excitement of 1892." The small size of the gold made recovery difficult and unprofitable, but the presence of supposedly rich accumulations captured the imagination of many persons who expended large sums of money erecting and supplying plants for recovery of fine gold. Actual recovery has been small; no production has been reported since 1941.

Source: Unknown. The small size of the gold indicates transportation from a distant upstream source. Gregory (1938, p. 108), however, suggests that the Triassic and Jurassic rocks exposed in that area, said to contain gold averaging 8.5¢ per cubic yard, might have been the source of some placer gold, also very fine and derived from an unknown source.

Literature:

Baker, 1936: Quotes Gregory (1917); Miser (1924); no placer activity in 1928.

Butler and others, 1920: Quotes Gregory's description (1917) of placers.

Engineering and Mining Journal, 1892: Reports history of discovery of San Juan River placers; reports exploration but no production from placers.

―――― 1893b: Describes end of placer excitement; reports numbers of men leaving area; reports lack of gold in placers.

Gregory, 1917: History; location of placer-mining operations; size of gold; placer-mining problems.

―――― 1938: Early history; production of Nephi claim; location of placer deposits; source.

Haskell and Sayner, 1956: Reports results of survey and sampling of placers along San Juan River and Colorado River to Lees Ferry.

Miser, 1924: Locates some deposits; brief history; production estimates.

8. LOWER COLORADO RIVER PLACERS

Location: Along the Colorado River from the confluence with the San Juan River (San Juan and Kane Counties, Utah) south to Lees Ferry (Coconino County, Ariz.).

Topographic maps: All 15-minute quadrangles—Navajo Mountain, Cummings Mesa, Gunsight Butte, Lees Ferry (Ariz.).

Geologic map: Cooley, Harshbarger, Akers, and Hardt, 1969, Geologic map of the Navajo and Hopi Indian Reservations (pl. 1, sheet 1), scale 1:125,000.

Access: Placer localities are accessible by boat north of the dam and by U.S. Highway 89 to Lees Ferry south of the dam.

Extent: Some placer gold was found at Klondike, Mescan, and Wright bars on the lower Colorado River. These localities are in the main part of Lake Powell, just upstream from the Glen Canyon Dam, and have been inundated by the rising waters behind the dam. Very fine gold particles occur in shales of the Triassic(?) and Triassic Moenkopi Formation and the Triassic Chinle Formation in the vicinity of the Paria River at Lees Ferry, Ariz., and in silts derived from the erosion of these beds.

Production history: No recorded production. The placer deposits at Klondike bar were reportedly productive. The gold-bearing shales in the Lees Ferry area were the cause of excitement and ambitious recovery schemes during the period 1910–13, but the very fine size of the particles and the isolation of the area precluded successful mining. The actual gold content averaged only about 5¢ per ton of rock material.

Source: Unknown.

Literature:
Gregory and Moore, 1931: Results of survey and sampling of placers lower Colorado River to Lees Ferry.
Hunt and others, 1953: Locates placer bars; history.
Lawson, 1913: Gold content in shales and sandstones at Paria River.

9. BLUE MOUNTAIN DISTRICT

Location: South flank of the Abajo Mountains, T. 34 S., R. 22 E. (unsurveyed, in the Manti-La Sal National Forest).

Topographic maps: Monticello and Mount Linnaeus 15-minute quadrangles.

Geologic map: Witkind, 1964, Geology of the Abajo Mountains area, (pl. 1).

Access: Light-duty roads lead west from Monticello to various points within the Abajo Mountains; dirt roads branching from the main roads lead to Johnson and Recapture Creeks.

Extent: Small placer deposits of unknown extent occur along Johnson and Recapture Creeks, which drain the south flanks of the Abajo Mountains and extend south to the mesas near Blanding. No description of the gold-bearing gravels has been found.

Production history: Placer gold was found in Johnson Creek in 1893 during the "Bluff excitement" along the San Juan River. The finding of placer gold led to the location of small gold mines within the mountains. Early production from the placers is unknown but presumably was small. Between 1933 and 1935 small amounts of placer gold were recovered from the area. Fifty-one ounces of placer gold was recovered in 1933 by a small sluicing plant (unlocated) in Doozit Canyon using water from Johnson Creek. Crawford (1933) examined the gold from this area and found the particles to average about 0.7 milligram in weight. He states that high gold-bearing bench gravels from Blanding Mesa were the source of the more concentrated placers below.

Source: The placer gold was derived from erosion of small gold veins in the stocks and laccolith margins of the Abajo Mountains. Erosion apparently has concentrated gold in both creeks and older mesa gravels where reconcentration allowed further creek accumulation.

Literature:
Butler and others, 1920: Discovery of placer gold; production information.
Crawford, 1933: Describes size and concentration of placer gold found in gravels of Blanding Mesa.
Engineering and Mining Journal, 1897b: Brief description of mining area; size of gold found in Johnson Creek.
Gregory, 1938: Discovery of placer gold.
Ritzma and Doelling, 1969: Sketch map (fig. 2) locates placer deposits in Abajo Mountains.

U S. Bureau of Mines, 1934: Placer-mining operations; production.

Witkind, 1964: Notes presence of placer gold; minor production; lode source.

UINTAH COUNTY

10. GREEN RIVER PLACERS

Location: Along the Green River from Flaming Gorge Reservoir downstream to the Horseshoe Bend, in particular, T. 5 S., Rs. 23 and 24 E.; T. 6 S., R. 21 E.

Topographic maps: All 7½-minute quadrangles—Dutch John, Goslin Mountain, Split Mountain, Jensen, Vernal SE, Brennan Basin.

Geologic maps:

Hansen, 1965, Geologic map and sections of the Flaming Gorge area (pl. 1), scale 1:48,000.

Kinney, 1955, Geologic map and structure sections of the Uinta River-Brush Creek area (pl. 1), scale 1:62,500; map of erosion surfaces and late Cenozoic deposits (pl. 3), scale 1:62,500.

Untermann and Untermann, 1964, Geologic map of Uintah County, scale 1:125,000.

Access: From Vernal, the placers along the Green River are accessible by various paved and dirt roads.

Extent: Placers were mined at numerous places along the Green River where fine gold had collected in river bars and in some mesa gravels. The most actively mined or prospected deposits were those at the Horseshoe Bend, at Jensen, and below Split Mountain Gorge.

Horseshoe Bend is south of Vernal in Tps. 6 and 7, R. 21 E. (Vernal SE and Brennan Basin quadrangles). The placers are on the inner side of the large oxbow curve made by the river (sec. 2, T. 7 S , R. 21 E.) and reportedly extend a mile in length and a quarter of a mile in width.

Large quantities of gravels occur along the Green River near Jensen and were worked for the gold content at the Jensen Bridge (sec. 21, T. 5 S., R. 23 E., Jensen quadrangle). Gravels dredged in 1910 reportedly averaged 60¢ to $1.75 per cubic yard.

Placer deposits exist in the bend of the Green River near the mouth of Cub Creek (sec. 5, T. 5 S., R. 24 E., Split Mountain quadrangle). Placer gold is reported to occur in Daggett County along the Red Canyon part of the Green River in sec. 10, T. 2 N., R. 22 E. (Dutch John quadrangle) and in secs. 16 and 17, T. 2 N., R. 23 E. (Goslin Mountain quadrangle). The gold from these localities is said to be fine flour gold, according to R. G. Pruitt (in Hansen, 1965, p. 183–184); no placer production is reported.

Production history: The placers along the Green River have produced small amounts of placer gold yearly from 1905 to 1942; production before that time is unknown. The gold in these gravels is very fine and has been difficult to recover. Dredge operations at the Horseshoe Bend, Jensen

Bridge, and Split Mountain localities were apparently unsuccessful; the largest annual recovery was 334 ounces in 1941. Three ounces of placer gold credited to the Green River in Grand County in 1932 is included in Uintah County production.

Source: The source of the placer gold is not well known; Uintah County is not noted for gold ore deposits. Kinney (1955, p. 162) suggests that some of the placer gold may have been derived from the Precambrian rocks of the Red Creek Quartzite in Browns Park, northwestern Colorado, and the Precambrian Uinta Mountain Group. The Uinta Mountain Group is extensively exposed in the Red Canyon part of the Green River near the Flaming Gorge Dam and may have been the source for the gold reported in that area.

Literature:

Butler and others, 1920: Location of placer-mining operations; distribution of gold in gravels; accessory minerals; gold values.

Gardner and Johnson, 1935: Horseshoe Bend—placer-mining activity and techniques in 1932; Horseshoe Bar—placer-mining activity and techniques in 1932.

Hansen, 1965: Location; mining history; gold values in placers along the Green River in Red Canyon (from R. G. Pruitt, U.S. Bur. Land Management, written commun., 1960).

Kinney, 1955: Past placer-mining activity; source.

Mining Review, 1910a: Placer-mining activity with dredge at Jensen; average gold values of gravels; locates other claims north of Jensen.

Pruitt, 1961: Locates two placer-mining operations; grades of gravel; economic significance.

Ritzma, 1959: Notes presence of "flour gold" in Brown Park and Red Canyon.

OTHER PLACER DISTRICTS
EMERY COUNTY
11. EMERY DISTRICT

Placer gold from Emery County in 1916 (locations unknown) was probably mined in the Emery or Lost Springs district (T. 19 S., Rs. 11 and 12 E.), the only district in the county where gold has been found in lode deposits Alternatively, the placer gold might have been recovered from gravels along the Green River, which forms the boundary between Emery County and Grand County, for placer gold was recovered from the Green River in 1932 and credited to Grand County by the U S. Bureau of Mines.

Literature:

Bullock, 1967: Locates mining districts in county.

JUAB COUNTY
12. DETROIT (DRUM) DISTRICT

Placer gold was recovered from unknown placers in the Detroit (Drum) district in the Drum Mountains (T. 14 S., R. 11 W.) in 1932. The ores in

the district contain gold and silver in sulfide minerals in replacement fissure deposits and probably were the source of the placer gold.
Literature:
Butler and others, 1920.

PIUTE COUNTY
13. GOLD MOUNTAIN DISTRICT

Placer gold was recovered below the outcrop of the Annie Laurie vein, which is near the head of Mill Creek on the north flank of Signal Peak in the Tushar Mountains (sec. 11, T. 27 S., R. 5 W., Delano Peak 15-minute quadrangle). No record of placer production has been found for this district in this century; the small amount of gold recovered was probably found before 1900. The valuable ore deposits are gold-silver in a residue of manganese oxides and quartz in the upper parts of veins.
Literature:
  Butler and others, 1920: Notes occurrence of placer gold near Annie Laurie vein.
  Callaghan and Parker, 1962: Describes lode deposits in Gold Mountain district.

14. OHIO (MOUNT BALDY) DISTRICT

Placer gold was recovered in 1868 near the mouth of Pine Gulch Creek in Bullion Canyon, on the northeast flank of the Tushar Range (sec. 34, T. 27 S., R. 4½ W., Delano Peak 15-minute quadrangle). Gold deposits occur in veins along Bullion Canyon and have been mined where the sulfides have been oxidized and leached. Other placers reportedly occur below the Deer Trail mine (sec. 12, T. 28 S., R. 4 W.). No recorded placer production has been found for this district, and the placers were certainly very small.
Literature:
  Butler and others, 1920: Notes occurrence of placer gold in Bullion Canyon.
  Callaghan and Parker, 1962: Describes lode deposits in Ohio district.
  Mining World, 1904: Notes plans to develop placers below Deer Trail mine.

SEVIER COUNTY
15. DISTRICT UNKNOWN

One ounce of placer gold was recovered from an unknown district in Sevier County in 1967.

TOOELE COUNTY
16. CAMP FLOYD (MERCUR) DISTRICT

The first mining claim in the Camp Floyd district was a placer claim made by L. Greeley, April 20, 1870; the placers were not profitable because of the paucity of gold and the shortage of water. The district, at the south end of the Oquirrh Mountains (Tps. 5 and 6 S., Rs. 3 and 4 W.), is the third largest gold mining district in the State. The gold is very fine grained.

Literature:
Gemmell, 1897.
Gilluly, 1932.

17. CLIFTON (GOLD HILL) DISTRICT

Placer gold reportedly was discovered in 1858 at Gold Hill in the Deep Creek Mountains (Tps. 8 and 9 S., R. 18 W.); hostile Indians are said to have driven away the early prospectors, and little work was done. There is no recorded placer production for the district.

Literature:
Rohfling, 1924: History.

WAYNE COUNTY

18. COLORADO RIVER PLACERS

Placer gold was recovered from unlocated deposits in Wayne County in 1910, 1922, and 1924. Production records indicate that placer gold of fineness 906 was recovered from high gravel bars of the Colorado River. The Cataract Canyon part of the Colorado River form the boundary between Wayne and San Juan Counties for a distance of about 4 miles. I do not know whether the placer gold was recovered from this part of the river or from other parts that lie in Garfield and San Juan Counties.

Literature:
U.S. Geological Survey, 1910, 1922, 1924.

GOLD PRODUCTION FROM PLACER DEPOSITS

Utah ranks 14th in the United States (10th in the western continental States) in placer gold production. The U.S. Bureau of Mines (1967, p. 15) estimates that 75,000 troy ounces of placer gold was produced in Utah from 1792 to 1964. This amount corresponds to the estimate of placer gold produced from the placers in the Bingham district before 1900. Since other placers were worked in the State before 1900, and 5,405 ounces of placer gold was produced between 1904 and 1968, I estimate the total placer gold production at about 85,000 ounces for the State. Table 1 gives the available information on placer-gold production.

The Bingham district placers were the most productive in the State. Nearly all placer gold mined from Bingham Canyon was recovered before 1900. During the 20th century, the placers along the Green River in Uintah County and the Colorado River in Grand County have been the most productive. Most of the gold has been recovered by small-scale mining techniques—usually sluicing along the riverbanks and sand bars, but some larger operations were attempted, such as the dredges at three localities along the Green River and near Dewey along the Colorado River. The fine gold dis-

TABLE 1.—*Utah placer gold production, in ounces*

Map locality (pl. 1)	County and placer district	Estimated production: Discovery to 1900	Recorded production (data from U.S. Bur. Mines)			Total recorded production, 1904-68	Total estimated production
			1904-33	1934-42	1943-68		
	Major districts						
	Garfield:						
1	Colorado River	>500	401	159	0	560	>1,200
2	Imperial	>150	105	87	5	197	≈400
	Grand:						
3	Colorado River	Unknown	548	571	12	1,131	≈1,500
4	La Sal	0	142	161	1	304	≈500
	Millard:						
5	House Range	0	21	247	58	326	≈400
	Salt Lake:						
6	Bingham	75,000	19	15	0	34	>75,000
	San Juan:						
7	San Juan River	Unknown	231	14	0	245	>300
8	Lower Colorado River	Unknown	0	0	0	0	≈100
9	Blue Mountains (Abajo)	Unknown	51	8	0	59	≈100
	Uintah:						
10	Green River	Unknown	715	567	0	1,282	>1,500
	Other placer districts						
	Emery:						
11	Emery	0	5	0	0	5	5
	Juab:						
12	Detroit (Drum)	0	5	0	0	5	5
	Piute:						
13	Gold Mountain	Unknown	0	0	0	0	Unknown
14	Ohio (Mount Baldy)	Unknown	0	0	0	0	Unknown
	Sevier:						
15	District unknown	0	0	0	1	1	1
	Tooele:						
16	Camp Floyd (Mercur)	Unknown	0	0	0	0	Unknown
17	Clifton (Gold Hill)	Unknown	0	0	0	0	Unknown
	Wayne:						
18	Colorado River	0	80	0	0	80	80
	Total	>75,650	2,323	1,829	77	4,229	≈85,000
	Undistributed to districts		1,175	0	1		
	State total	75,650	3,498	1,829	78	4,229	85,000

[1] Includes production from San Juan County placers.

tributed in the sands and gravels along the major rivers in Utah presented many recovery problems. The burlap-covered sluice was the most successful modification to standard placer-mining techniques.

SUMMARY

The placer deposits of Utah occur in two distinct environments—placers found adjacent to, and derived from, gold, silver, and base-metal deposits and placers found in major rivers and derived from unknown, distant source areas.

PLACERS ASSOCIATED WITH KNOWN LODE DEPOSITS

The major placer district in Utah, the Bingham district, contains two major types of ore deposits, both of which contain gold. The prophyry-type copper ores contain only minor amounts of gold, but byproduct recovery currently places Bingham as the third largest gold-producing district in the country. The nonporphyry ores consist of fissure veins and replacement deposits mined primarily for lead-silver-zinc; in oxidized portions of some of these deposits, free gold has been concentrated in honeycombed siliceous gangue. Both types of ore are related to the intrusion of the Bingham stock. Most of the placer gold recovered from gravels in Bingham Canyon and side gulches was derived from the nonporphyry ores, but some placer gold recovered from gravels near the Bingham stock is thought to have been derived from the porphyry ores. The Bingham district has undergone four stages of dissection followed by relatively short periods of deposition, apparently throughout a long period in the Quaternary; the placer gold found in bench and creek gravels is the product of this rapid erosion and subsequent deposition.

Other productive placers that are found adjacent to known lode deposits include those in the Imperial district (Garfield County), La Sal district (Grand County), House Range (Millard County), and Blue Mountains (San Juan County). The ore deposits in these districts are all associated with stocks of presumed Tertiary age, and excepting the unknown source in the House Range, consist of small bodies of gold, silver, and base-metal ore; gold was deposited with debris eroded from the stocks and enclosing rocks to form small placer deposits. The gold-bearing source in the House Range has not been recognized, but the placer gold is found in arkosic sands and gravels derived from the granitic intrusive rock which forms the range and from which the gold was presumably derived. The placers in the La Sal district are found in glacial deposits of Pleistocene age; all other placers are found in accumulations of fanglomerates and gravels of Quaternary age.

Placer gold was reported to occur in four of the major lode-gold districts in Utah; but no placer production was recorded from these districts, and the amount of placer gold must have been exceedingly small. The reasons for the insignificance of detrital gold derived from these important lode deposits are diverse. In the Gold Mountain district, the gold was concentrated in a leached residue of manganese oxide and quartz, which precluded the formation of free gold for erosion. Some native gold occurs in the oxidized fissure veins; in the Ohio and Clifton districts, apparently the gold is very fine and therefore did not form workable placers. The Camp Floyd district contains gold in replacement ores, but there, too, the gold is very fine grained and did not form workable placers.

Bergendahl (1964, p. 87), in summarizing the occurrence of placer deposits, states,

In part the relative lack of placer deposits in comparison to other western areas reflects the kind of mineralization found in Utah's mining districts. Free gold is sparsely present in a few ores, but in most is intimately associated with metallic sulfides and would be released in a very finely divided state during weathering.

The occurrence of typically fine-grained gold in most of the major lode-gold districts in the State accounts for the fact that no placer gold has been recovered from these districts.

RIVER PLACERS

Placer gold is found in sand and gravel bars at many localities along the Colorado River and its major tributaries, the San Juan, Dolores, and Green Rivers. The question of ultimate source of the gold found along these rivers has been an enigma to those who have studied and worked the river placers.

So far as I know, no detailed studies of the origin of the gold by trace element analysis, sediment transport, associated heavy mineral suites, and so on, have been made on any of the river placers in Utah; therefore, the origin of the placer gold remains unknown.

The rivers in the southeastern part of the State drain large parts of the Colorado Plateaus province, an area underlain by thick deposits of sedimentary rocks of Mississippian through Cretaceous age. Placer gold has been recovered from river bars along the length of the Colorado River from the Dolores River south to Lees Ferry, Ariz., a distance of about 200 miles, and from the mouth of Montezuma Creek on the San Juan River west to the junction with the Colorado River, a distance of about 100 miles. In this large area, only three lode-gold districts are known, all of them minor: the La Sal district, 16 miles south of the Dolores River and 12 miles east of the Colorado River; the Imperial district, 24 miles west of the Colorado River; and the Blue Mountains district, 36 miles north of the San Juan River. Drainage from these districts reaches the Colorado and San Juan Rivers and could have supplied some of the gold found along these rivers, but it cannot account for the presence of gold placers upstream. Part of the gold may have been derived from lode deposits many miles upstream (for example, the San Juan Mountains in Colorado) and transported from this distant source.

Another possible source for the placer gold in southeastern Utah is the sedimentary rocks exposed. Gold reportedly occurs in sandstones adjacent to the San Juan River and in sandstones in the Henry Mountains, west of the Colorado River. The Triassic Shinarump Member of the Chinle Formation in the vicinity of Paria (Kane County) contains gold averaging 8.5 cents per cubic yard, but attempts to mine the deposits were a failure. The source of the gold within these Paleozoic and Mesozoic rocks is unknown, but the deposits apparently are low-grade, widely disseminated fossil gold placers.

The placers along the Green River are in part of the Rocky Mountain province and the northern part of the Colorado Plateaus province in the northeastern part of Utah. The geology of the area through which the major part of the Green River drains is considerably different from that of the area drained by the Colorado and San Juan Rivers. The Uinta Mountains are composed of a core of relatively unmetamorphosed sedimentary rocks of late Precambrian age flanked by sedimentary rocks of Paleozoic and Mesozoic age. The source of the placer gold in the Green River is unknown, but it may be in the Precambrian rocks of the Uinta Mountains and the Red Creek Quartzite in Browns Park.

BIBLIOGRAPHY

LITERATURE REFERENCES

Baker, A. A., 1936, Geology of the Monument Valley-Navajo Mountain region, San Juan County, Utah: U.S. Geol. Survey Bull. 865, 106 p.

Bergendahl, M. H., 1964, Gold, *in* Mineral and water resources of Utah [Report of the United States Geological Survey in cooperation with Utah Geological and Mineralogical Survey and the Utah Water and Power Board]: U.S. 88th Cong., 2d sess., Senate Comm. Interior and Insular Affairs, Comm. Print, p. 83–89.

Discusses reasons for comparative lack of placer deposits in State.

Boutwell, J. M., 1902, Ore deposits of Bingham, Utah: U.S. Geol. Survey Bull. 213, p. 105–122.

Abstract of Prof. Paper 38.

——— 1905, Economic geology of the Bingham mining district: U.S. Geol. Survey Prof. Paper 38, 413 p.

Bullock, K. C., 1967, Minerals of Utah: Utah Geol. and Mineralog. Survey Bull. 76, 237 p.

Burchard, H. C., 1882, Report of the Director of the Mint upon the statistics of the production of the precious metals in the United States (for the year 1881): Washington, U.S. Bur. Mint, 765 p. [Utah, p. 237–248].

Butler, B. S., Loughlin, G. F., Heikes, V. C., and others, 1920, The ore deposits of Utah: U.S. Geol. Survey Prof. Paper 111, 672 p.

History, location, and mining history of placers in State.

Callaghan, Eugene, and Parker, R. L., 1962, Geology of the Delano Peak quadrangle, Utah: U.S. Geol. Survey Geol. Quad. Map GQ-153, scale 1:62,500.

Crawford, A. L., 1933, An application of microscopy for evaluating the gold in certain Utah placers [abs.]: Utah Acad. Sci. Proc., v. 10, p. 57.

Crawford, A. L., and Buranek, A. M., 1944, Amazon stone, a new variety of feldspar for Utah—with notes on the laccolithic character of the House Range intrusive: Utah Acad. Sci. Proc., 1941–43, v. 19–20, p. 125–127.

Dane, C. H., 1935, Geology of the Salt Valley anticline and adjacent areas, Grand County, Utah: U.S. Geol. Survey Bull. 863, 184 p.

Engineering and Mining Journal, 1892, General mining news—Arizona [San Juan River placers]: Eng. and Mining Jour., v. 54, p. 613.
—— 1893a, General mining news—Utah [Imperial district]: Eng. and Mining Jour., v. 56, p. 122.
—— 1893b, General mining news—Utah [San Juan River]: Eng. and Mining Jour., v. 55, p. 40.
—— 1897a, General mining news—Utah [Imperial district]: Eng. and Mining Jour., v. 63, p. 493.
—— 1897b, The Blue Mountains in Utah: Eng. and Mining Jour., v. 63, p. 574.
—— 1936, News of the industry—placer mining in the Colorado River, Grand County, Utah: Eng. and Mining Jour., v. 137, p. 581.
Gardner, E. D., and Johnson, C. H., 1935, Placer mining in the western United States, Part 3: U.S. Bur. Mines Inf. Circ. 6788, 81 p.
Gehman, H. M., 1958, Notch Peak intrusive, Millard County, Utah: Utah Geol. and Mineralog. Survey Bull. 62, 50 p.
Gemmell, R. C., 1897, The Camp Floyd mining district and the Mercur mines, Utah: Eng. and Mining Jour., v. 63, p. 403–404; 427–428.
Gilluly, James, 1932, Geology and ore deposits of the Stockton and Fairfield quadrangles, Utah: U.S. Geol. Survey Prof. Paper 173, 171 p.
Gregory, H. E., 1917, Geology of the Navajo country, a reconnaissance of parts of Arizona, New Mexico, and Utah: U.S. Geol. Survey Prof. Paper 93, 161 p.
—— 1938, The San Juan country, a geographic and geologic reconnaissance of southeastern Utah: U.S. Geol. Survey Prof. Paper 188, 123 p.
Gregory, H. E., and Moore, R. C., 1931, The Kaiparowits region, a geographic and geologic reconnaissance of parts of Utah and Arizona: U.S. Geol. Survey Prof. Paper 164, 161 p.
Hammond, E. D., 1961, History of mining in the Bingham district, Utah *in* Utah Geol. Soc., Guidebook to the geology of Utah: no. 16, p. 120–129.
Hanks, K. L., 1962, Geology of the central House Range area, Millard County, Utah: Brigham Young Univ. Geology Studies, v. 9, pt. 2, p. 115–136.
Hansen, W. R., 1965, Geology of the Flaming Gorge area, Utah-Colorado-Wyoming: U.S. Geol. Survey Prof. Paper 490, 196 p.
Haskell, H. S., and Sayner, D. B., 1956, Gold, *in* Kiersch, G. A., Metalliferous minerals and mineral fuels, v. 1 *of* Mineral resources, Navajo-Hopi Indian Reservations, Arizona-Utah: Tucson, Univ. Arizona Press, p. 33–38.
Hill, J. M., 1913, Notes on the northern La Sal Mountains, Grand County, Utah: U.S. Geol. Survey Bull. 530, p. 99–118.
Hunt, C. B., 1958, Structural and igneous geology of the La Sal Mountains, Utah: U.S. Geol. Survey Prof. Paper 294–I, p. 305–364.
Hunt, C. B., Averitt, Paul, and Miller, R. L., 1953, Geology and geography of the Henry Mountains region, Utah: U.S. Geol. Survey Prof. Paper 228, 234 p.
Kinney, D. M., 1955, Geology of the Uinta River-Brush Creek area, Duchesne and Uintah Counties, Utah: U.S. Geol. Survey Bull. 1007, 185 p.
Lakes, Arthur, 1908, Geology and economics of Rio San Juan, Utah: Mining World, v. 28, p. 761–762.
—— 1910, The origin of flour gold: Mining Sci., v. 61, p. 52–53.
Lawson, A. C., 1913, The gold of the Shinarump at Paria: Econ. Geology, v. 8, p. 434–446.
Maguire, Don, 1899, Bingham Canyon mines: Mines and Minerals, v. 19, p. 377–378.
Mardirosian, C. A., 1966, Mining districts and mineral deposits of Utah: Salt Lake City, Utah, Charles A. Mardirosian, privately printed.
  Map (scale 1:750,000) shows location of mining districts; table notes districts where placer gold was produced.

Mining Journal, 1929, Mining in the Mountain States—Utah [Imperial district]; Mining Jour. [Phoenix, Ariz.], v. 13, no. 7, p. 45.

——— 1938, Concentrates from the western states—Utah [House Range]: Mining Jour. [Phoenix, Ariz.], v. 22, no. 14, p. 34.

Mining Reporter, 1899, World's richest placering [Colorado River placers]: Mining Reporter, v. 39, no. 3, p. 16.

Mining Review, 1910a, Placer mining on Green River; Mining Rev. [Salt Lake City, Utah], v. 12(?), (July 30, 1910), no. 8, p. 21.

——— 1910b, The Wilson Mesa: Mining Rev. [Salt Lake City, Utah], v. 12, no. 3, p. 22.

Mining Science, 1910, La Sal vein and placer gold: Mining Sci., v. 61, p. 84–85.

Mining World, 1904, Mining News from busy mining camps—Utah [Ohio district]: Mining World, v. 21, p. 436.

Miser, H. D., 1924, The San Juan Canyon, southeastern Utah: U.S. Geol. Survey Water-Supply Paper 538, 80 p.

Moore, W. J., Lanphere, M. A., and Obradovich, J. D., 1968, Chronology of intrusion, volcanism, and ore deposition at Bingham, Utah: Econ. Geology, v. 63, no. 6, p. 612–621.

Mullens, T. E., 1960, Geology of the Clay Hills Area, San Juan County, Utah: U.S. Geol. Survey Bull. 1087–H, p. 259–333.

Powell, D. K., 1959, The geology of southern House Range, Millard County, Utah: Brigham Young Univ. Geology Studies, v. 6, no. 1, 49 p.

Pruitt, R. G., Jr., 1961, The mineral resources of Uintah County: Utah Geol. and Mineral Survey Bull. 71, 101 p.

Raymond, R. W., 1870, Statistics of mines and mining in the States and Territories west of the Rocky Mountains (for the year 1869): Washington, U.S. Treasury Dept., 805 p. [Utah, p. 319–321].

——— 1872, Statistics of mines and mining in the States and Territories west of the Rocky Mountains for the year 1870: Washington, U.S. Treasury Dept., 566 p. [Utah, p. 218–223].

——— 1873, Statistics of mines and mining in the States and Territories west of the Rocky Mountains being the fourth annual report (for the year 1871): Washington, U.S. Treasury Dept., 566 p. [Utah, p. 300–330].

——— 1874, Statistics of mines and mining in the States and Territories west of the Rocky Mountains being the sixth annual report (for the year 1873): Washington, U.S. Treasury Dept., 585 p. [Utah, p. 255–283].

——— 1877, Statistics of mines and mining in the States and Territories west of the Rocky Mountains being the eighth annual report (for the year 1875): Washington, U.S. Treasury Dept., 519 p. [Utah, p. 267–281].

Ritzma, H. R., 1959, Geologic atlas of Utah, Daggett County: Utah Geol. and Mineralog. Survey Bull. 66, 111 p.

Ritzma, H. R., and Doelling, H. H., 1969, Mineral Resources, San Juan County, Utah, and adjacent areas—Part 1, Petroleum, potash, groundwater, and miscellaneous minerals: Utah Geol. and Mineralog. Survey Spec. Studies 24, 125 p.

Summarizes general location, character, and value of placer deposits in the Colorado, Green, San Juan, and Dolores Rivers and in the La Sal and Abajo Mountains.

Rohfling, D. P., 1924, Ore deposits of the Deep Creek Range, Utah: Salt Lake Mining Rev., v. 25, no. 24, p. 11–15, 18–19.

Rubright, R. D., and Hart, O. J., 1968, Non-porphyry ores of the Bingham district, Utah, in Ridge, J. D., ed., Ore deposits of the United States, 1933–1967, v. 1: New York, Am. Inst. Mining, Metall. and Petroleum Engineers, Inc., p. 886–907.

BIBLIOGRAPHY

Thaden, R. E., Trites, A. F., Jr., and Finnell, T. L., 1964, Geology and ore deposits of the White Canyon area, San Juan and Garfield Counties, Utah: U.S. Geol. Survey Bull. 1125, 166 p.

U.S. Bureau of Mines, 1925–34, Mineral resources of the United States [annual volumes, 1924–31]: Washington, U.S. Govt. Printing Office.

―――― 1933–68, Minerals Yearbook [annual volumes, 1932–65]: Washington, U.S. Govt. Printing Office.

Information relating to placers cited in text is referenced by year of pertinent volume.

―――― 1967, Production potential of known gold deposits in the United States: U.S. Bur. Mines Inf. Circ. 8331, 24 p.

Gives estimates of total placer gold production in troy ounces.

U.S. Geological Survey, 1883–1924, Mineral resources of the United States [annual volumes, 1882–1923]: Washington, U.S. Govt. Printing Office.

―――― 1896–1900, Annual reports [17th through 21st, 1895–1900]: Washington, U.S. Govt. Printing Office.

Information relating to placers cited in text is referenced by year of pertinent volume.

Utah Geological and Mineralogical Survey, 1966, Gold placers in Utah—a compilation: Utah Geol. and Mineralog. Survey Circ. 47, 29 p.

A compilation based on publications of U.S. Geological Survey and Utah Geological and Mineralogical Survey; extensively quotes placer descriptions from these publications.

Witkind, I. J., 1964, Geology of the Abajo Mountains area, San Juan County, Utah: U.S. Geol. Survey Prof. Paper 453, 110 p.

GEOLOGIC MAP REFERENCES

[References keyed by number to districts given in text]

Boutwell, J. M., 1905, Economic geology of the Bingham mining district: U.S. Geol. Survey Prof. Paper 38, pl. 1; pl. 2 (shows distribution of placer gold deposits).
No. 6.

Cooley, M. E., Harshbarger, J. W., Akers, J. P., and Hardt, W. F., 1969, Regional hydrogeology of the Navajo and Hopi Indian Reservations, Arizona, New Mexico, and Utah: U.S. Geol. Survey Prof. Paper 521–A, pl. 1(sheets 1,3,5), scale 1:125,000.
Nos. 7, 8.

Hanks, K. L., 1962, Geology of the central House Range area, Millard County, Utah: Brigham Young Univ., Geology Studies, v. 9, pt. 2, geologic map.
No. 5.

Hansen, W. R., 1965, Geology of the Flaming Gorge area, Utah-Colorado-Wyoming: U.S. Geol. Survey Prof. Paper 490, pl. 1.
No. 10.

Hunt, C. B., 1958, Structural and igneous geology of the La Sal Mountains, Utah: U.S. Geol. Survey Prof. Paper 294–I, pl. 40 (bedrock geology).
No. 4.

Hunt, C. B., Averitt, Paul, and Miller, R. L., 1953, Geology and geography of the Henry Mountains region, Utah: U.S. Geol. Survey Prof. Paper 228, pl. 1.
Nos. 1, 2.

Kinney, D. M., 1955, Geology of the Uinta River-Brush Creek area, Duchesne and Uintah Counties, Utah: U.S. Geol. Survey Bull. 1007, pls. 1, 3.
No. 10.

Richmond, G. M., 1962, Quaternary stratigraphy of the La Sal Mountains, Utah: U.S. Geol. Survey Prof. Paper 324, pl. 1 (Quaternary geology).
No. 4.

Untermann, G. E., and Untermann, B. R., 1964, Geology of Uintah County: Utah Geol. and Mineralog. Survey Bull. 72.
No. 10.

Utah Geological Society, 1961, Geology of the Bingham mining district and northern Oquirrh Mountains, D. R. Cook, ed.: Utah Geol. Soc. Guidebook to the geology of Utah, no. 16, pl. 2.
No. 6.

Williams, P. L., 1964, Geology, structure, and uranium deposits of the Moab quadrangle, Colorado and Utah: U.S. Geol. Survey Misc. Geol. Inv. Map I-360, scale 1:250,000.
No. 3.

Witkind, I. J., 1964, Geology of the Abajo Mountains area, San Juan County, Utah: U.S. Geol. Survey Prof. Paper 453, pl. 1.
No. 9.